# NURTURING FRIENDSHIPS

# NURTURING FRIENDSHIPS

*Building Meaningful Connections*

AVERY NIGHTINGALE

# CONTENTS

| | | |
|---|---|---|
| 1 | Introduction | 1 |
| 2 | Understanding Friendships | 3 |
| 3 | Benefits of Meaningful Connections | 5 |
| 4 | Developing Authentic Relationships | 7 |
| 5 | Communication Skills for Building Friendships | 9 |
| 6 | Building Trust and Loyalty | 10 |
| 7 | Overcoming Challenges in Friendships | 11 |
| 8 | Supporting Friends in Difficult Times | 13 |
| 9 | Maintaining Healthy Boundaries | 15 |
| 10 | Creating Shared Experiences | 17 |
| 11 | Celebrating Milestones and Achievements | 19 |
| 12 | Balancing Individuality and Togetherness | 21 |
| 13 | Resolving Conflicts in Friendships | 23 |
| 14 | Cultivating Empathy and Understanding | 24 |
| 15 | Nurturing Long-Distance Friendships | 26 |
| 16 | Building Friendships in Different Life Stages | 28 |

| | | |
|---|---|---|
| 17 | Fostering Friendships in the Workplace | 30 |
| 18 | Maintaining Friendships Through Technology | 32 |
| 19 | Sustaining Friendships in a Busy World | 34 |
| 20 | The Role of Vulnerability in Friendships | 36 |
| 21 | Supporting Mental Health in Friendships | 38 |
| 22 | Embracing Diversity in Friendships | 40 |
| 23 | The Importance of Listening in Friendships | 42 |
| 24 | Giving and Receiving Emotional Support | 44 |
| 25 | Building Friendships in a New Community | 46 |
| 26 | Balancing Online and Offline Connections | 48 |
| 27 | Recognizing Toxic Friendships | 50 |
| 28 | Letting Go of Unhealthy Relationships | 52 |
| 29 | Cultivating Gratitude in Friendships | 54 |
| 30 | The Impact of Friendships on Well-being | 56 |
| 31 | Conclusion | 58 |

Copyright © 2024 by Avery Nightingale

All rights reserved. No part of this book may be reproduced in any manner whatsoever without written permission except in the case of brief quotations embodied in critical articles and reviews.

First Printing, 2024

# CHAPTER 1

# Introduction

Components for the cultivation of culture. This aims to be a comprehensive essay, the first of its kind in the study of personal and cultural development, which investigates various human qualities and conditions from the perspective of the gene. While much of evolutionary psychology stands upon examination of the adaptive problems of our hunter-gatherer forebears, thus establishing the ceiling upon which genetic evolution operated, little attention has been paid to the psychology of the intervening epoch and its impact upon the genome. This essay adapts the concepts of Trimorphism and Fund of Social Capital as a means of discussing three forms of cognitive and emotional processing prevalent in society today, as well as the potential for genetic adaptation to its circumstance. At a primitive level, culture exists as basic infrastructure - a developing repository of technology, traditions, procedures, and ethics designed to enhance the quality of life. As genetics exist as the program outlining the development and function of an organism, it can be said that culture is not simply an ephemeral event in history, but a lasting aggravator of genetic evolution, and the alliance of genes and culture has shaped human conditions across the globe. In order to understand cultural impacts upon genetic adaptation to types of problems, we must first understand specific cultural problems and their relation to mental phenomena in light of human survival and reproduction.

# CHAPTER 2

# Understanding Friendships

The text for section "2. Understanding Friendships" is: Peplau and Perlman (1982) define friendship as a voluntary relationship in which people enjoy each other's company. Friendship occurs frequently between people and associations are formed through companionship and sharing. The friendships field is continuously focused on the nature of the particular social exchange, the characteristics of individuals who become friends, and the formation process. Social exchange theorists stress that interaction is based on the balance of benefits and rewards between the two people. Thus, relationships are seen as a series of transactions that are designed to produce the greatest profit possible. Over time, should rewards or benefits become scarce or not delivered at all, relationships will terminate. According to this view, friendships are chosen in accordance with the need for valuable resources. These resources can be tangible or intangible items that can be exchanged at a very low cost or for free. One reason why the resource theory is seen as quite influential is that it highlights an essential aspect in both friendship and close relationships: exchange and equity. This theory is seen as an adequate explanation of couples who have just moved in together or friends who are currently flatting. A different idea of social exchange has been introduced by Fehr (2009). Fehr's view on cooperation suggests that when people engage in cooperative activities, interactions, or relationships, they generally have a conditional cooperative stance. This means they are predisposed to cooperating but are also influenced in their cooperation by prior action or other parties involved. Fehr shares a similar belief in that cooperative relationships are a form of implicit agreement to continue cooperation as long as an individual's expectations are not disappointed. Friends are likely to expect certain actions from their friends based on how cooperative they think their friend should be to sustain the relationship. High satisfactory levels (described as the difference between joint benefits and the next best alternative) will result in cooperative expectations being fulfilled and a continuation in the relationship. If at any stage a friend has expectations disappointed, expectancy discontinuity occurs and must be repaired to save the relationship. This act of relationship repair is essential to the continuation of a friendship and asks for forgiveness.

# CHAPTER 3

# Benefits of Meaningful Connections

Friends also offer a counter perspective; they can offer advice and opinions that open our eyes to different ways of thinking. They can alert us to any self-destructive behavior. This again aids problem-solving, and a friend who aims to increase the emotional strength of their friend can also encourage mental activity in attempts to change unhealthy behavior. This type of support, involving encouragement for someone to take better care of themselves, is called persuasive support.

A benefit of friendship that is crucial to healthy mental well-being is that of validation support. This occurs when friends offer emotional support in terms of listening to problems and showing concern. The benefit is that discussing emotional problems increases understanding of the problems, decreases the emotional burden, and most importantly, it helps with problem-solving by getting emotional issues out of the way.

The support we get from friends may also encourage us to attain a healthy lifestyle. The same University of Virginia study found that those with friends were more likely to use effective problem-focused coping strategies and active problem-solving.

Friendship also makes it more likely that we deal with certain problems and worries. A study conducted at the University of Virginia in 1988 found that when adults were presented with a distasteful task, those with friends were more likely to feel contempt and try to take action to change the situation. This involved rumination, a coping strategy which is passive and has been connected to depression. In simple terms, those with friends were less likely to dwell on their problems and more likely to look for a solution.

Another benefit of friendship is reduced stress, which can also adversely affect health. According to psychologist Debra Umberson, having friendships can undo the effects of stress on both psychological and biological reactions to stressful events. The idea is that having a friend to turn to when things get rough helps a person cope by lending them moral support, talking about problems, or even doing a distracting activity. The friendship activates behavioral and hormonal mechanisms that protect against the effects of stress.

Meaningful friendships are about something more, as are the benefits of developing and maintaining them. Friends play a crucial role in helping us manage the health of our minds and bodies. That is the seminal finding of a trio of reports on friendship recently published in the journal Developmental Review. The study claims that a lack of friendships can be hazardous to health, even for a middle-class adult just below social class. People who maintain vital friendship networks at work and all levels of society have a lower rate of mortality, and of course, friendlessness is not the only factor in mortality. This fact proves that friends are like a tonic.

## CHAPTER 4

# Developing Authentic Relationships

At this level, there is a greater commitment to authentic sharing and an increasingly clear understanding of the give and take in the relationship. And, as mentioned previously, friendships have a chance to become more intimate over time. An authentic friendship is built on affirmation and support. Over time, friends provide each other with so much support and affirmation that a strong sense of trust and respect emerges. Being able to discuss personal concerns and receiving understanding and encouragement builds closeness among friends.

Just as there are levels of involvement in shaping nurturing relationships, there are also levels of friendship. While different experts may categorize types of friendships somewhat differently, they generally agree that friendships can begin with casual friends and move to close friends or Best Friends Forever, as school-aged girls like to call it. True friendships typically begin and remain at the close friend level, with gradual deepening over time.

True friendship is based on mutual acceptance and understanding, and it blossoms when friends share a view of life and common values. When we develop an authentic friendship with someone, we typically experience a deeper sense of intimacy, and we feel we can share important parts of ourselves. While we may have many acquaintances - people who are friendly and like to chat - acquaintances are not the same as friends. Friendship usually implies a relationship that is genuine, and as we bring more of our true, authentic selves into the relationship, we can increase the level of intimacy.

# CHAPTER 5

# Communication Skills for Building Friendships

B. Self-assertion skills: 1. Ensure you are always clear in what you are asking or saying. 2. Ensure your nonverbal communication matches your verbal communication. 3. Maintain eye contact and use the person's name to convey that you have their undivided attention. 4. Ask for feedback from others regarding the conversation.

1. Guidelines for communicating and initiating friendships: 1. Self-disclosure - Share some information about yourself. Start with superficial information and gradually move on to more personal topics. 2. Do not start with very personal information because if it is rejected, it may have negative implications for further friendship. 3. Be lighthearted and fun, as this is more attractive to others. 4. Be positive in your interactions with others and try to ask questions about them to show genuine interest. 5. Take steps to arrange more regular meetings instead of relying on chance encounters.

## CHAPTER 6

# Building Trust and Loyalty

One of the most essential actions for building trust is to be honest. In an ideal world, honesty would be straightforward and easy, but being truly honest with someone, especially when telling our true feelings about them or something that they have done, can be incredibly difficult. Honesty is not without its risks, especially if an apology is involved, but it is the bedrock of any healthy relationship and is the only real way to resolve conflicting issues, thereby increasing understanding between the two parties.

Of course, this is easier said than done. Trust, like a bridge, has to be built slowly over time. Loyalty, like trust, is a vital ingredient for all friendships, but can also be very fragile. We all want to know that a friend is dependable, that they will be there for us when we need them the most. Just as importantly, we all want to know that our friends are loyal to us behind our back, especially when we are not around. There are many things that you can do to build trust and loyalty in your friendships, here are a few of the most essential.

# CHAPTER 7

# Overcoming Challenges in Friendships

Resilience is key in overcoming these obstacles. The lonely person must be prepared to accept that setbacks are par for the course, that people are not perfect, and that misunderstandings and hurts in friendships are not the end of the world. They must remain steadfast in the face of their own self-sabotaging behavior and push through the discomfort. It is helpful in this process to have someone who understands and is patient and forgiving, although it is difficult to forge friendships like these immediately. Often, the lonely person will need to rely upon professional help or a support group to help them through this process. With time and various trials and errors, the person will slowly start to develop more effective ways of communicating and resolving conflicts and will begin to feel a greater sense of confidence in their ability to sustain a longer-term friendship.

Friendship, even at its best, often brings up all kinds of complications and difficulties. A deep friendship will, almost inevitably, be tested by disagreements, by an unthoughtful word or ill-placed remark, and even by outright betrayal. For a recovering lonely person, even the idea of going through this with a friend can be enough for them to back out. They are too used to a life of pain and hurt, and very often they find the security of loneliness to be the better option. In fact, many will unwittingly self-sabotage a friendship that gets too close. They do not feel worthy of a good friend and feel it is only a matter of time before the friend finds them out to be as boring and as unlikeable as they believe themselves to be. All these kinds of thoughts are self-fulfilling prophecies, and it is common that the person will create distance in the relationship and eventually feel relieved when it comes to an end. Step by step and little by little, the person needs to change their pattern of behavior and thinking in relationships so as to start overcoming these obstacles and building stronger connections.

# CHAPTER 8

# Supporting Friends in Difficult Times

First, it is important to recognize when a friend is having a tough time. This is not always as obvious as it may seem. Some individuals may suddenly change their behavior or attitude. Others may seem more tired and run down or more quick-tempered. This could be for a number of reasons, and a change in behavior should never be judged. The best thing to do is gently encourage your friend to talk about it. Remember that not everyone is comfortable discussing their emotions, and they may not want to talk to you about it. Let them know that you are there to listen if they want to talk, but do not pester them. The fact that you have shown concern and are there to listen is often enough to help them through. Another sign that a person is troubled is if their school work starts to suffer or they struggle at work. In this case, your support may be to help them through a tough time without actually discussing what the problem is. Pick up on any invitations to spend time with you or do something fun and ensure that they know you are there to help. Often pleasant distractions can make people forget their problems and can work as great therapy.

**CHAPTER 9**

# Maintaining Healthy Boundaries

In any kind of relationship, there is often a blurring between the roles of friend and the roles of listener and problem-solver. This is not necessarily unhealthy, and many friendships thrive on helping each other through tough times. However, it is important to be aware of not being too heavily reliant on one another. There is a difference between being supportive and being a crutch. It is important to cultivate your own coping resources so that you are not dependent on someone else. This can be done by seeking professional help from a counselor or therapist. Keep an open mind and be willing to try new things. Your friend might not always be able to provide an answer for you, and it is healthier for both of you to not get frustrated by this. Remember to stay balanced in a given role of helping and being helped. It is a common trap in many friendships where one person is always the giver, and the other is always the receiver. This can lead to feelings of resentment and a lack of personal growth.

The feeling of connection and unity between friends can, at times, undermine the need for autonomy and individuality within relationships. As you become closer to your friends, it is important to remember that you are still an individual with your own thoughts, beliefs, and feelings. Rediscover who you are and what you want in your life. It can be easy to become caught up in the lives of others, but it is essential to maintain a strong sense of self. Continue to grow and develop as a person. Be mindful of your own needs. Remember that you have a right to say no to requests from your friends, and it is okay to express your own needs. You don't have to do everything together. Do not be afraid of speaking honestly and assertively with your friends. Good friends will be able to respect your honesty, and it will only serve to strengthen the relationship in the long run. Understand the importance of time alone and don't be afraid to let your friends know that. Time alone helps to maintain your individuality and strengthens your sense of independence. It also allows you to bring fresh energy into relationships.

# CHAPTER 10

# Creating Shared Experiences

To take a relationship to another level, shared experiences must occur. These experiences may be planned, such as going on a trip or to a concert. They simply may be activities a person does not want to do alone, like going to the grocery store. Often times it is those "spur of the moment" occurrences that create the best lasting memories. Victor defines shared experiences as "those moments when you and your friend(s) do the same thing together." Research has shown that it doesn't matter what the experience actually is, but what actually creates the lasting memories and bonds are the interactions between people during the experience. It is during these experiences that people can truly get to know one another, find out if they have similar interests, and if their personalities are compatible. Another point to remember is that it is okay to share negative experiences with a person or group of friends. It is naive to think that everything shared between you and another person will be positive. In fact, lasting relationships will have their ups and downs, and it is the downs that help us to learn more about each other and ourselves. Going through rough times with a person and helping each other out can create strong bonds of trust and compassion.

# CHAPTER 11

# Celebrating Milestones and Achievements

Children will want to celebrate their achievements with family and friends. One idea is to have a commemorative event such as a dinner party following the child's participation in a significant activity, event, or competition. The child may wish to send invitations to their friends and have the opportunity to share their experience and show any resulting trophies, ribbons, or medals. Parents who take a significant interest in their child's social and emotional development may choose to write a friendship resume of the child's achievements. This could be letters or diary entries documenting specific events and the child's thoughts and feelings. These memories can be reflected upon in the future and provide evidence of the child's growth and learning in the area of friendship.

Children will have an immense sense of pride and accomplishment when they have milestones and achievements celebrated by others. Milestones provide an opportunity to recognize a child's growth and development, while achievements are a measurable result that further the child's progress. Nurturing friendships from the youngest age, parents can encourage simple milestones. For instance, forming a first meaningful friendship may be cause for celebration. The growth of the friendship can be monitored and when the children encounter a dispute, parents may determine an appropriate resolution, offering praise when the children have resolved the issue effectively. Milestones and achievements in friendship are important in building a child's self-esteem and confidence. By offering praise and support, children determine that friendship is something of value to them and well worth future effort.

# CHAPTER 12

# Balancing Individuality and Togetherness

On the other hand, togetherness is about bonding with others and working as a team so that everyone involved feels a sense of belonging and personal value. It is essentially about an exchange of selves that results in the collective self, involving shared identities, shared understanding, and interdependence. The pressures of togetherness become very real when we have friends in need, a family to support, a partner, or an important work team.

In order to juggle better, we need to understand the forces that pull us towards individuality and togetherness, as well as set some clear and reasonable goals. Individuality is about who we think we are and what we can become. It is about self-awareness, self-reliance, and personal growth. It is a muscle that needs to be flexed often in order to develop; hence, some time alone is essential in order to not lose touch with oneself. For some people, individual goals may be very high, i.e., "I want to be a leader in my field" or "I want to be successful enough to provide a good lifestyle for my family." For others, it may be as simple as "I want to find out who I really am" or "I want to be happy." All these individuals will find time alone attractive and valuable.

We are multitasking creatures capable of doing one act while thinking of another. When people look at our companions, they explain, "I don't know how she does it." In the fast-paced, individual-oriented modern world, we are pulled in a variety of directions by people and commitments, including our friends and acquaintances, our family, and our work. Maintaining friendships as well as spending time alone can be a "juggling act," and feeling drained, confused, or divided is not uncommon.

# CHAPTER 13

# Resolving Conflicts in Friendships

Firstly, you need to approach the person with whom you have a conflict in a non-confrontational way. Tell them that you value their friendship and that you want to resolve the issue that you currently have. Make sure that you have the conversation in a private place where you will not be disturbed, which will prevent either person from feeling embarrassed or getting up and walking away before the issue is resolved. This step is crucial as the way and the environment in which a conflict is resolved have a large effect on the outcome.

Conflicts are inevitable in all friendships; they can be a source of irritation or a tool used to build a stronger relationship. No matter how they start, all conflicts in friendships have a chance to be resolved and can serve to build a deeper understanding of both yourself and the friend involved. Resolving conflicts honestly and coming out with a stronger relationship is an important set of skills that can serve you well in later life and in your other relationships. Next time you and a friend are in an argument, try these simple steps to resolve the issue.

# CHAPTER 14

# Cultivating Empathy and Understanding

A third type of social cognition can be used to build friendship or several different types. An example of this would be using definitive logic and reasoning with a person who is sorrowful. While the intent would be to raise the person's spirits, it also improves the ability of understanding that will lead to enhanced empathy in future similar situations. This is a type of supportive behavior and is often used to establish a better understanding of what the supportive person is trying to express.

Empathy is a much easier concept to explain. It is the ability to respond to another person's feelings or emotions with an appropriate emotion. Also, it is important to note that empathy does not mean pity. Pity is feeling sorry for another person's misfortune and is much less constructive in nature. A simple example of empathy would be if you saw someone burn their hand, you would wince in expectation of their pain. This is quite a bit different from the reaction of hysterical laughter one would find from a child seeing another child scold themselves for the first time. A lack of empathy is due to a lack of maturity and/or experience in similar situations.

How do empathy and understanding foster better relationships? In truth, the better you understand the emotions and thoughts (cognition) of another person, the more likely it is you will be able to react in a way that facilitates the desired objective for both of you. Not to mention the fact that understanding begets better problem solving and aids in a time of crisis. Empathy helps to develop understanding and, in turn, understanding administers harmony.

**CHAPTER 15**

# Nurturing Long-Distance Friendships

For those who feel that the majority of their friends are so far away, it is quite possible to maintain a long-distance friendship. Whether it is kids who have moved to another town or friends met across the country or the globe, through the use of regular mail, e-mail, phone calls, and visits, children and adults can maintain a strong friendship. For some busy families, getting a new pet or taking care of a friend's pet can have the added benefit of providing another routine approach to staying in touch. Long-distance friendships can be nurtured through many of the same activities that sustain local friendships. These activities often serve as a way to remove the barrier for scheduling get-togethers, because at any time the friends can feel connected. Ongoing examples include sharing jokes, stories, or interesting news, continuing a book or hobby together which has a satisfying conclusion or finished product, and playing games with set turns, complete with post-game analysis. When sharing any of these experiences, it is important to give the friend your undivided attention when interacting. Remember to be understanding and considerate of the friend's situation and feelings. It is important to recognize that people and relationships can change over time. View it as a chance to have a new experience, and fear not, for you can still keep in touch with your old friend. Finally, long-distance friends can feel a sense of togetherness by fostering an attachment to prior shared experiences and being optimistic about creating new memories when they meet again.

**CHAPTER 16**

# Building Friendships in Different Life Stages

According to Adams and Berner, there are gender and generational differences in how friends are viewed. Men are more likely to have common activities such as sports or work-related responsibilities. Their friendships may be "insulated" against changes in life problems and not always emotionally supportive. It is said that male friendships are a shield against depression and loneliness. Post-generation X has witnessed a shift in ideals for male friendships, but many are still traditional in nature and resistant to change. Female friendships are generally longer and encourage intimate conversation. Adams and Berner recommend that best friends are found to fill the emotional need and encourage mutual reinforcement between friends of emotional support and understanding. Having this understanding of differing types of functional friendships can help individuals to identify their own friendships and evaluate what they mean to them.

With milestones in the lifespan, such as marriage or parenthood, there comes a threat to the relationship of long-standing friends.

Adams and Berner say that a study of married women found that after marriage, women had fewer friends, and life satisfaction was strongest for those who maintained more friendships. This indicates that a change in lifestyle, such as marriage and having children, portends the need to make more of an effort to maintain friendships and social contacts.

Relationships are something that we all need, and when it comes to friendship, it is just as important. "Maintaining friends is one of the most important sources of social support for adults". This statement is discussed by Adams and Berner in a textbook on understanding psychology. In the textbook, they state that a study of middle-aged women found that those who had "close long-term friends were less depressed and lonely than those without them." This evidence proves just how important it is to ensure that friendships are maintained throughout the lifespan.

## CHAPTER 17

# Fostering Friendships in the Workplace

Whites and Asians are more likely than Latinos and Blacks to initiate workplace friendships. This is often due to the feeling of being an 'out group' member based on minority status within the company. For these people, it is the duty of the potential friend and the organization to make them feel welcome and that they truly belong as part of the team.

Some managerial guidance is appropriate, as currently few people have leadership experience for creating an environment and guiding the processes that lead to close and constructive relationships among co-workers. High job involvement and a supportive organizational climate bode well for the development of workplace friendships. Personalities make a difference as well. In addition to having the time and opportunities to be with co-workers, those who are emotionally stable, not prone to negative moods, and low on introversion are more likely to build workplace friendships. Just because a person is extraverted though, does not guarantee that they are apt to build friendships at work.

Fostering workplace friendships can be tricky. You don't want to appear unprofessional, but spending time with colleagues in a relaxed setting helps create the bonds that can make a job more satisfying and the workload seem lighter. Just as friendships formed in high school or college, workplace friendships are based on common interests, history, proximity, and shared emotional experiences. The strongest workplace friendships often are based on shared values and experiences in taking on challenges and overcoming obstacles in the work environment. Strong friendships lead to stronger teams, so it is to your advantage to help employees forge them.

**CHAPTER 18**

# Maintaining Friendships Through Technology

The widespread use of the internet and email in contemporary society has given rise to an entirely new method of maintaining and forming friendships. I am almost certain that each one of you reading this has a Facebook account. It may not be your cup of tea, but with one out of every thirteen people on earth having an account, there is a good chance that many of your friendships were initiated or maintained through Facebook. In this incredibly fast-paced society, we are always looking for ways to get any task done more efficiently. Unfortunately, this quite often includes the way we communicate. Social networking sites and email may cut down on the time it takes to say hello to a friend, but they ultimately delete that personal interaction between the two. As I mentioned before, the quality of communication is far more significant than the quantity. Steps should be taken to limit the time spent communicating through such sites and more time need be allocated to more personal forms of communication. Though emails and these sites can be useful in initially setting up plans between friends, conversations through such media tend to be very open-ended and plans are often forgotten. With the invention of social networking sites such as Facebook, Myspace, and Twitter, it is becoming easier and easier to know what is going on in the lives of friends. This can be both good and bad. Knowing when a friend's birthday is, or when they have a big test or event coming up allows you to show that you care by acknowledging such milestones. On the other hand, relying on information gained through social networking sites or the like does not compare to hearing such information directly from a friend. This is the same as with certain events happening in a friend's life. While knowing about such events can be advantageous, it is more meaningful to simply hear these things from the friend himself. The bottom line is, while it is very convenient to have information about friends at our fingertips, we must not forget that the personal touch is what truly means the most.

## CHAPTER 19

# Sustaining Friendships in a Busy World

In today's fast-paced society, scheduling conflicts are conflicting friendships everywhere. You may have promising plans to meet a friend for lunch, only to find that a last-minute appointment has been made with your academic advisor. Or perhaps, as you excitedly prepare for a night out, you receive a phone call that your cousin is in town for one night and you are her only available babysitter. Although disappointments are inevitable, it is essential that you follow through with your original plans whenever possible. Do not cancel time with a friend to engage in another activity that you could do at any time; putting off your immediate plans has become a habit of procrastination. Often, our friends get the message that they are not our first priority. Make them feel valued by placing importance on your commitments to them. Hopefully, they will understand when your situation is vice versa.

With your current friends, make an effort to reach out and stay in touch. Offer a time that you are available to talk, or plan an activity out together. It will make a difference. You can also allocate time specifically for catching up with friends, and use the remaining time for study, work, and alone time. Remember also, spending time with good friends energizes you and increases your overall happiness.

Start with truly wanting to sustain your friendships. If it is not a priority for you, it will not happen. You must value the importance of your relationships and want them to continue. Put the effort into fostering your friendship, rather than just going along with the stream. You will need to be committed to devoting time and energy into nurturing your relationships. It is easy to let connections fade, but far harder to redevelop a friendship.

# CHAPTER 20

# The Role of Vulnerability in Friendships

Whether vulnerability is discussed in the context of an interaction between friends, a marriage, or a business organization, the concept is often met with negative connotations and associated with some form of weakness. In their responses, participants in the present study commonly used negative descriptors when discussing vulnerability as it relates to expressing difficult emotions or aspects of the self. One respondent compared it to "going out on a limb," and with regards to communicating vulnerability with a friend, another respondent stated, "either we do and things will work out and be deeper, or things will not go well and we will not ever understand each other more than we do now." With responses like these, it is evident that there is some level of discomfort or even fear when it comes to revealing one's true self and a hesitation to let down the walls that may prevent one from doing so. Despite these negative connotations, our respondents showed an evolved and more positive view of vulnerability in comparison with previous research on the topic, and they all expressed that it was an essential component to forming close relationships with friends. I agree with these beliefs, and while discussing my findings with my own friends, it was easy to come up with examples of how vulnerability has propelled our friendships to new levels of closeness. So while it may have taken a difficult conversation or a leap of faith to express an aspect of our true selves, we all agreed that the end result was a deeper understanding and knowledge of one another and feeling more at ease in our friendships due to the increased closeness.

Simply defined, vulnerability is the act of showing and expressing your true and authentic self. It is a concept that was constantly brought up when respondents were asked to define the essential qualities of a close friendship. Respondents often stated that when meeting someone, the other person would have to "open up" to them or "let them in." This "opening up" was repeatedly described as the sharing of one's true thoughts and feelings and being oneself around the other person. This true self that is revealed to another is the product of vulnerability, and it is what forms the deeper connections and close relationships that these respondents were so full of.

# CHAPTER 21

# Supporting Mental Health in Friendships

Despite the positive implications of good friendships on mental health, the absence of supportive friends is a risk factor for depression. This shows that while good friendships can enhance mental health, those that are negative or absent can be detrimental. It is important that people who are experiencing mental health difficulties are able to communicate this to their friends. Self-disclosure has been found to be positively linked with emotional support and negatively related to feelings of alienation, and it is suggested that having the opportunity to disclose problems may facilitate greater support from friends. This indicates that friends are more likely to provide support if they are aware it is needed and suggests that clear communication is essential in allowing friends to provide effective support.

It is an unfortunate reality that one in four individuals will be affected by mental health disorders within their lifetime. Due to the strong prevalence of mental health issues, it is essential that all friendships provide the necessary support to foster positive mental health and resilience. Peer support has been strongly linked with recovery from mental health difficulties and friendship satisfaction. Having supportive friends has been a protective factor in the prevention of suicide, the highest cause of death in young people from the UK. Friends have been found to be an essential outlet for releasing negative emotions and instrumental in the maintenance of good mental health. A recent study found that among a group of young adults, the presence of social support protected highly stressed individuals from developing depression. This demonstrated the healing value of friends and indicates the importance of social support in combatting stress.

## CHAPTER 22

# Embracing Diversity in Friendships

Tolerance. True understanding can lead to tolerance. It's normal to feel a certain degree of discomfort when we are immersed in a culture different from our own. But for the most part, people are people, and we all share the same basic wants and needs in life. If we can learn to accept the differences and even appreciate the unique qualities of people different from ourselves, we can have a very rich experience in life.

Overcoming Prejudices. If you are open to understanding cultures different from your own, you will be less likely to fall into patterns of unwitting prejudice. How often do we judge people by our own cultural norms? People from some Asian cultures, for example, might find it difficult to look you in the eye, as a sign of respect. By understanding and accepting this cultural difference, we are less likely to regard the person as "shifty" or "untrustworthy".

Understanding Diversity. One key to building relationships among people of different cultures (or between different age groups, or economic classes) is truly getting to know each other. Ask questions, listen, observe. Remember, it's what we have in common that builds the initial bond of the relationship.

Your friends don't have to be your mirror image. In fact, friends who are different from you can offer you a wealth of knowledge and experience that you might not attain on your own.

## CHAPTER 23

# The Importance of Listening in Friendships

This final analysis is very important and may be the bane of many friendships when done poorly. Have you ever had a friend cut you off mid-story with a story of their own? Ever been asked a question that was answered with a completely different subject? Many people are uncomfortable listening to deep or emotional issues. They try to avoid the discomfort by quickly providing a solution or attempting to distract the friend. While it is well-intentioned, it effectively cuts off the person's attempt to work through their own issue. Conversely, sometimes the listener simply becomes lost and when the storyteller notices, they get frustrated and feel as if they are burdening the listener. These issues are at the root of many friendship misunderstandings and fade-outs.

Learning to be a good listener will allow you to help your friends sort through their own confusions. As our friends talk, we help them think through their issues by sorting out their story and expressing the emotions behind it. We ask open-ended questions and provide reflective comments that encourage deeper thoughts and feelings. We empathize with them, rather than commiserating. We develop a better understanding of our friends and actually increase our own cognitive skills. Active listening involves hearing between the lines (interpretation) and giving feedback on what was heard (evaluation). In the end, good listening skills help us do more than we could have by simply talking.

Sometimes, after a day spent with a friend, you find yourself asking, "What did we really talk about?" The answer is probably not much. Although we live in an impatient society, one of the kindest things embodied in and from our friendships is our willingness to tell and retell our stories. However, storytelling is only half of the equation in communication. The other half, listening, is just as important.

# CHAPTER 24

# Giving and Receiving Emotional Support

It is interesting to note gender differences for this aspect of friendship. Female and male same-sex pairs are more similar in the way that they give and receive emotional support than cross-gender pairs. Majors (1996) identified that women were more likely than men to give empathic responses to their friends' sadness, and women were more likely to seek such responses when they themselves were sad. Boergers et al. (1995) reported that women were more satisfied with same-sex friendships than were men. A reason for this could be that women are getting the emotional support that they desire, whereas it is not as forthcoming in cross-gender friendships. This is an overgeneralization, of course, there are some men who excel in giving emotional support. At the end of the day, it differs between individuals.

The value of emotional support is priceless. It can be anything from a smile when one is feeling low, a pat on the back, a friend's presence in a time of crisis, or just knowing that someone is there to share one's joy and sadness. Emotional support contributes a great deal to the success of a friendship. It is easier to weather the tough times when one has a caring friend and the joy in life is multiplied when it is shared with others. Friends who are always on the receiving end, however, may find this an unequal relationship and will feel uncomfortable about it. It is important that both parties have a chance to give and receive emotional support.

## CHAPTER 25

# Building Friendships in a New Community

• Language classes, gyms, coffee shops, churches, places of employment are all good spots to look for potential friends. The key is to put yourself in contact with people with similar interests to you, and then it is a matter of developing the relationship.

• Volunteer. This activity can be a quick way to meet people and to find out what types of people are in the community. Events in the wider community often use volunteers and various programs often are looking for board members.

• Get to know the neighbors. If you live in an apartment, make an effort to meet people in the same building. Once you see a familiar face a few times, start to chat with the person and look for common ground.

Relocation can be a lonely experience, even for those who have moved many times. Just knowing how to find potential friends can be daunting. Sometimes the difference between meeting potential friends is as simple as being somewhere where you have the opportunity to meet people, i.e. attending a church of the same denomination, an expat community, a newcomers club. Consider these suggestions when trying to build friendships in a new community.

# CHAPTER 26

# Balancing Online and Offline Connections

When thinking about virtual friendships, potential problems can blur the solid advantages. Online networks are hurt by issues of trust and identification. When a person's true identity is not capable of being known, it might be tricky to expand genuine friendship. It is also hard to build trust in a friendship when the communication is solely online. Seeing if a person is really who they say they are and judging their level of honesty is simpler in offline communication in face to face situations. Cyberstalking is another problem that might end up in highly linked persons finding difficulty in trusting each other. Perpetually checking up on someone's online activity might lead to a lack of self-assurance in the friendship and it could generate undesirable tension between the two parties. This problem might be extra troublesome if a person is suspected of attempting to escape an online friendship by bifurcating off and building a new virtual friendship with somebody else. Lastly, it is feasible that an online friendship might conflict with a person's offline relationships and resources. If this occurs too much, the individual might feel isolated from their offline companions and it might potentially harm their personal life. This whole situation might generate feelings of regret and lead to dissolution of the web-based friendship.

**CHAPTER 27**

# Recognizing Toxic Friendships

Sometimes being around certain friends can cause an emotional drain. Perhaps you have a certain friend that wants to hang out with you, but when you do, you feel hesitant about it. Even if it is at a subconscious level, he may be causing emotional stress and proving more hassle than worth. Emotional drain can be caused by a friend that is relying too much on your own emotional support, especially if they are needier emotionally than you are. The depth of their personal problems and emotions may be difficult for you to handle if you are not in a position to help them or deal with it effectively for them.

Oftentimes, a single event can be justifiable, but a repeated pattern of behavior is hard to ignore. Look at the friend's actions and behavior in the recent past. Have they been supportive to you or have they put you down? A friend who is offhand with their insults is still an insult. Words are difficult to forget and hurt more than sticks and stones. Remember, if it's justifiable but it really isn't worth the pain received, often the argument to justify an insult can be worse.

One of the first steps of learning how to nurture friendships is being able to recognize a toxic one. Toxic friendships are ones that make you feel bad about yourself, bring emotional stress or pain to your individual life, and don't benefit you. Identifying the negative friendship can be difficult due to feelings of obligation and guilt, but recognizing the friendship is not aiding you positively is the first step to freeing yourself from it. Usually, the warning signs appear subconsciously and you know when you have a negative encounter with an individual, but suppress it when it is a friend. Here are some more clear warning signs to figure out if a friend is toxic.

## CHAPTER 28

# Letting Go of Unhealthy Relationships

There are many kinds of relationships and friendships that are not all created equal. When determining whether or not to keep or end a relationship, you might find it helpful to divide the relationship in question into different categories such as acquaintance, friend, and close friend. Considering that many people who you have communicated with have made some kind of impact on your life, dissect whether or not the impact was significant enough to warrant a continued relationship with said person. Keep in mind the type of person that you know and whether or not they have certain negative habits, attitudes or outlooks on life that constantly clash with or bring you down from your own. This may include people who are constantly negative, critical, unsupportive or even those who are too needy and constantly drain your energy. All of these considerations will help you to recognize whether or not the relationship is one that is mutually beneficial to both parties involved. If it is determined that the relationship in question is one that would be best ended, it is important to make a clear break. This is not to say that all bridges should be burnt and ties severed but there is no sense in drawing out an unhealthy relationship. It is possible that you may encounter the person in question at another point in life where the relationship can take a turn for the better but until that time, it is a good idea to move on.

Begin by examining your relationships and determining which ones no longer deserve your time and energy. This requires taking a long, hard look at yourself and your role in the demise of the relationship. It's important to remember that many times a relationship is not ended abruptly or without cause. So take stock of what the relationship has meant to you and whether or not it has improved your life or made it more difficult. If the relationship is one that truly enhances your life or brings you vital experiences, try to work on the negative aspects that are making it unhealthy with the friend in question. If the relationship is not one that has made a positive impact on you, you will want to categorize what type of relationship it is and plan to break it off.

**CHAPTER 29**

# Cultivating Gratitude in Friendships

Emmons' research on the benefits of gratitude is relevant to friendship in that grateful people are more likely to acknowledge a kind act, reflect upon it, and often repay it. This is clearly important in the initiation phase of new friendships in which the goal is to discover common ground. A positive emotional connection creates openness and the free exchange of information. This often leads to an invitation to do something in the way of friendly behavior. An individual who holds a negative emotional connection with the potential friend will often miss, misinterpret, or avoid the friendly gesture. This will not foster an amicable relationship between the two.

In one study, Emmons found that people who kept gratitude journals on a weekly basis exercised more regularly, reported fewer physical symptoms, felt better about their lives as a whole, and were more optimistic about the upcoming week compared to those who recorded hassles or neutral life events. The benefits were clearer when compared to the control group that did not record anything. A related benefit was observed in the realm of personal goal attainment, with the participants in the gratitude condition progressing more rapidly toward their goals.

What's the perfect antidote to entitlement? Gratitude. It strengthens interpersonal bonds and enriches the lives of those who cultivate it. Robert Emmons, a researcher in the field of positive psychology, defines gratitude as "a felt sense of wonder, thankfulness, and appreciation for life." Although the vast majority of gratitude research has been conducted at the individual level, its relevance to the formation and maintenance of social bonds is clearly evident.

# CHAPTER 30

# The Impact of Friendships on Well-being

Experimental and cross-sectional studies demonstrate the detrimental effects of social isolation and lack of social support on health and health-related behaviors. In recent meta-analyses of clinical and epidemiological studies, social support was significantly related to better functional outcomes for various medical conditions. Cross-sectional and prospective longitudinal studies provide strong evidence that the prevalence of social support and the quality of social relationships are causally related to the incidence of mental and physical health problems. Several behavioral pathways have emerged as candidate mediators for the observed effects of social support on health and health-related behaviors. Despite support for the link between social support and health, underlying mechanisms and long-term effects remain unclear.

Social relationships are consistently tied to mental health and well-being. However, most research in the field emphasizes the consequences of negative social exchanges on psychological outcomes such as depression and anxiety. The idea that positive social relationships might contribute to psychological well-being has remained relatively unexplored. Considerable correlational evidence suggests the presence of a robust association between social support and mental health. Moreover, several national opinion surveys document a consistent decline in social participation in the United States over the last two decades. Despite converging evidence of the crucial importance of social relationships to mental health, questions remain regarding the processes by which social and psychological factors are related.

# CHAPTER 31

# Conclusion

An emphasis on the importance of social connections and trusting relationships has important implications for participatory models of intervention and capacity building. An understanding of intra-group and inter-group social capital can facilitate a process in which disenfranchised communities are able to identify and utilize resources to take control over and improve their health status. This is a sharp contrast to approaches that have often pathologized the poor and minority populations, focusing on identification of deficits and imposition of solutions from outside. Work which aims to build leadership and collective efficacy in disenfranchised communities can potentially empower individuals to advocate for their children and create environments that are more protective of maternal and child health.

Beyond the United States, we have begun to investigate social support and social capital as critical resources related to maternal and child health in developed and developing countries around the world. A few examples illustrate the potential ways in which understanding the social context can help to inform more effective program development. In Singapore, we are investigating how social support from friends and family influences maternal infant feeding decisions in a multi-ethnic population undergoing rapid social change. In the post-economic crisis climate in Argentina, we are examining the ways in which social capital can safeguard the continued access to resources for maternal and child health among the most disenfranchised and poorest sectors of society. An ongoing project with an international non-governmental organization, Nutrition International, aims to improve access and utilization of health and nutrition services among the poor in developing countries by constructing knowledge about specific ways to strengthen community social capital resources.

For instance, in developing and applying strategies to make the transition to motherhood, lower income women emphasized the importance of developing a trusted network of family, friends, and neighbors who could provide the kind of instrumental and emotional support necessary to care for young children. They described nurturing their relationships with providers and informal networks as a way to secure better treatment for their children. Middle income mothers cited the specific ways women come to understand, support and sustain each other in learning through the exchange of information, community building and skill building activities. Understanding social capital may also explain preliminary findings from a comparison of middle and lower income women in a study of breastfeeding and postpartum weight loss in North Carolina. Subtle intervention on part of health providers resulted in different resources accessed by two groups of women. Middle income women, who reported higher levels of social capital, were more likely to become linked to internet-based breastfeeding support.